EARTH SONG

◇

*by
Miriam
Kaplan*

Copyright © Miriam Kaplan
All rights reserved.

Printed in the United States of America.
First Printing 1988

ISBN 0-9620-233-1-0

Crystal Heart Press
Fallsburg, N.Y. 12733

Front cover photo by Bill Davis and Louie Kampouris
Front cover photo © Miriam Kaplan 1988
Typography by The TypePeople, Binghamton, NY

Dedication
To Bhagavan Nityananda

ACKNOWLEDGEMENTS

Special thanks to Ron Benenati, Franko Richmond, Joan Goldstein, Peg Bendet, Lynn McCulloch, and Judy Parsons for their assistance and support.

Also thanks to Bill Davis, of Taos, New Mexico, for the majority of photographs in the book and to Glenda Newell-Blake for the design of the book.

Preface

This book is meant to be experienced through the heart. Its basis is the sacredness of the earth and her place in the heavens. My own inner inspiration, inspiration from other writers, research concerning attributes of the earth, and interviews with people who are intimately connected to the earth make up the contents of this book.

Miriam Kaplan

Psalm 24

The earth is the Lord's, and the fullness thereof; The world, and they that dwell therein. For He hath founded it upon the seas, And established it upon the floods. Who shall ascend into the hills of the Lord? Or who shall stand in his holy place? He that has clean hands and a pure heart.

Earth Song
TABLE OF CONTENTS

CHAPTER 1. *The Relationship of the Earth to the Heavens*

The Earth — Our Home

The Law of Reciprocity

The Earth's Relationship to the Sun

The Earth and her Relationship in the Solar System

The Earth in Relation to Cosmic Events

Consecration of the Earth

Celebrations

Vision of the Heavens

Meditation on the Sun, Moon, and the Stars

Meditative Exercise — Alignment with the Earth and Heavens

CHAPTER 2. *Spirit of the Earth*

Gaea

Lessons from Nature

Nature Spirits and the Devic Kingdom

Presence of the Angels

Revelations from Nature

Horticulture and Telepathy

Crystals and Mining:
Interview with Gary Fleck

Caring for the Land

The Spirit of the Forests: The Sacred Tree

Impressions of the Amazon Rain Forest:
Interview with Rusty Crutcher

Shaman as Ecologist

The Body Physical: Establishing Harmony within The Body

Give the Earth Nourishment:
Interview with Don Juan Camargo Huaman

The Five Elements

Meditation — The Earth as a Body of Light

CHAPTER 3. ***The Grid of the Earth: Sacred Sites***

Healing with the Earth: Giving and Receiving Energy

Magnetic Flow in our Lives

Energy Currents in our Bodies

The Earth's Energy Currents

The Grid Pattern

Meditation on the Earth's Grid

Crystalline Grid and the Australian Aboriginals: *Interview with Lorraine Mathi Williams*

Sacred Sites

Approaching Sacred Sites

Holy Mountains

Mt. Kailasa

Sacred Sites of the Mayan: *Interview with Hunbatz Men*

A Prayer for the Earth

CHAPTER ONE

◇

The Relationship of the Earth to the Heavens

◇

The Earth — Our Home

The planet earth is many things to all of us. First of all, she is our home, she is where we live. Earth is also our mother, in that she provides us with food, with plants to heal us, with stones and crystals to teach us and to attune us. She provides us a place to be while we evolve, while we go through life's experiences to acquire knowledge and wisdom.

Many in the native traditions call her mother earth and this is a good name. For through her sustenance we are able to grow. If our bodies are thought to be a garb for the soul then our beautiful planet earth can be thought to be a raiment for our bodies.

The earth's relationship to the other planets in the solar system is an interesting one. Both polarities of existence, the negative and the positive, are present together. While living on this earth, and being sustained by her, we experience both pain and pleasure, gain and loss, givers and exploiters.

There are many other planets or planes of existence. Some planets are oriented towards pleasures, but not towards spiritual growth or evolution. Others are even more negative than the earth; yet others are very subtle, beautiful and spiritualized. On those planets, light beings exist, because the lessons that only duality teaches have been learned. Whatever plane of existence is appropriate for the evolving soul is where the soul manifests. We who are on earth are given a rare opportunity to evolve and to realize our divinity. Through the lessons, through the hardship, we are able to go beyond the forms that are presented to us and also to go beyond the various attachments to these forms. If we are lucky enough to meditate, to receive holy communion within, we can be taken to the most pure and holy realms while still being in this body.

The earth is our home, our dwelling place while we grow and recognize our inner light. But the earth is also a living organism, who herself is evolving. Her force field is a living one and while we are on her, we are in direct communication with her whether we are conscious of it or not.

The crisis that the earth is in, and at the same time humanity is in, necessitates greater consciousness on our part of the mother earth. As she serves us by presenting us with the very rare opportunity to evolve, we must begin to serve her by protecting her body through prayer and good actions.

The earth is said to be the planet of purification of the solar system. Yet she is known to be the most beautiful of all the planets in this solar system. Times of crisis such as these can be turned into times of truth and learning if we heed our inner voice and listen to what the earth is saying.

The Law of Reciprocity

All creation is founded on certain energy imprints, or spiritual laws. Adhering to these laws creates balance and harmony because we function within the way of creation. Order is perpetually maintained. Through the following of spiritual law comes renewal. When the order of a higher law is not followed, disorder and chaos manifest. For instance, in a disease such as cancer, individual cells no longer care about the good of the whole. Due to this lack of caring and of effective communication, the organism is eaten up and eventually loses its vitality. It is obvious that we cannot find a cure for cancer until we can self-correct this malfunctioning. Another extension of this symptom is in the pollutants in our environment that, in effect, cause cancer. A few individuals are out to get as much as they can with no regard for the environment or others. The pollutants they create are in the streams, the rivers, the very

circulatory system of the earth. Nuclear wastes are buried deep in the earth. Our forests, the skin of our planet, are being ripped out. And so the earth loses her protective aura.

However, the law of reciprocity keeps on functioning. As we continue to pollute the earth, with no regard for her well-being, we create disease that forces us to slowly die. There have been so many books about healing ourselves but we must not overlook the Law of Reciprocity. This Law is very simple: if we take, we must give back in return. If we only take from the earth and do not give back to her, we will suffer plagues and earth disturbances, such as droughts. If we can turn our hearing within and listen to the song that the earth is now singing, we will hear that our mother, who has fed us from birth, is suffering and is angry. This is our time to listen and to mature, and to learn the laws of creation that have been taught since ancient times. The very amazing thing about understanding and following the Law of Reciprocity is knowing the truth that as we give so we are given to. As we learn to give nourishment to the earth and to others, and as we learn to truly take nourishment from the earth, so other patterns in our life are affected. Rather than growing up, as many of us did, with expectations of fulfillment of our immediate desires and the desires of our culture, we learn instead to be grateful for what we are given and to give back with joy. We enter a continuous cycle of creativity. And we learn to recycle the gift of energy we receive. Through this activity our perception changes and we slowly begin to perceive our holiness.

Planets, as well as humans, change vibrational frequencies as they evolve. The earth is currently in the process of changing her vibrational frequency to a higher, lighter vibration. We, too, are in the process of changing our vibrational frequency and moving towards our true heritage which is to become beings of light. This means that our process is accelerated. Part of our inner process at this time, and into the next decade, is to adjust our physical and mental bodies to a higher vibrational state.

One way to do this is to meditate, to go within and to commune with the heart. This is our true home. There we derive the answers to our questions and quench our long thirst for Truth.

Crystals, and other stones, assist us in both stabilizing our energy field and adjusting our electrical impulses. Attunement to these higher vibrational frequencies that are flowing into us and through us now is what is required of us. This attunement is an inner awakening, and it is time for us to acknowledge the importance of receiving and transmitting light. We must be both participants and vehicles for this planetary awakening that is occurring. Part of this attunement as well is to attune to the earth, our physical home, and to begin to realize what the words "taking stewardship" mean.

The Earth's Relationship to the Sun

The sun is the life giver for the earth. The sun is the aspiration of the earth. The solar light force permeates the earth and gives incentive for the earth's evolutionary process or movement into light. Ancient cultures understood the link to the sun and built their structures in alignment with solar and lunar events. This understanding included the metaphysical conception of alignment with the force of light, the giver of all life.

The ancients' understanding, in cultures such as Stonehenge, Newgrange, Machu Picchu, did not separate their science from their spirituality. They did not negate one to give credence to another. Rather the universe was conceived as a living whole whose movements deeply affected us. Solar alignments were considered to be of key importance because they were known to occur within our own inner being. The inner and the outer corresponded to one another. The sun, through its alignment at

the solstice or the equinox, became, once again, beams of light. The earth in her sacredness absorbed this energy and issued forth this energy.

The ancients understood and demonstrated that this universe is indeed a mystical one, a magical one, and that the inner and the outer both influence one another and reflect one another. In acknowledging the sun in the alignment of their structures, they acknowledged the giver of all life.

The Earth and her Relationship in the Solar System

The earth is in relationship to all of the planets in the solar system and to other galaxies. This relationship is both on an inner level and an outer level. It exists in the physical universe and it exists on the subtle planes, or metaphysically. On the subtle levels sound and light are motivating forces in the universe. In sound there is total resonance to the sacred sound Om.

The earth has her own sound tone in the universe, and that sound tone resonates to all the other planets in his entire galaxy. In the same way, the other planets in the solar system and galaxy also resonate to their own tone in correspondence to each other and to the sun as well. This resonance creates a vibration where life unfolds. The earth's body resonates to a certain evolutionary octave. This too resonates with all of the other planets, both within our solar system and also in interconnecting galaxies. This means that all of life, all of consciousness, is connected.

In our solar system, the structure of consciousness revolves around that which gives the most life: in our vision, it is the sun.

The outer sun also corresponds to the inner sun, the giver of light and of life, and of knowledge.

On the physical level, the earth, because of her desecration by toxins, mining, and stripping of her trees, is resounding with a sound vibration of distress and of disease. This resonance is affecting other planets in our solar system and galaxies.

The Earth in Relation to Cosmic Events

Just as we have subtle bodies, so the earth also has subtle bodies. One of the functions of the subtle bodies of the earth (which is also linked to the grid of the earth) is to receive and to process cosmic events, both planetary and galactic, and to bring them gently into the earth's own energy field.

As the earth moves into a higher vibration, this continuous occurrence becomes all the more obvious and significant. The influence on the earth of planetary, solar, lunar, and galactic events was recognized in ancient cultures. Here structures were built, such as Stonehenge, that recorded these events. The structures were also aligned with the sun, the moon, and different planets.

The ancients understood that the earth was not an isolated body in space but was connected to the heavenly bodies. They knew the relationship of the earth to the heavenly bodies influences times of planting and harvesting, how and when ceremonies were to occur and the consciousness of humans. The earth and the heavens were not considered to be separate from man or woman but were rather a whole continuum of consciousness.

The ancient structures could be viewed as "tuning forks" for various cosmic events, partly due to their alignments and the numerical proportions in which they were built. They were built on power spots so that energy from the heavens could be received and distributed effectively.

Human beings also understood themselves to be vehicles for cosmic intergalactic events. Certain people were trained in a priestly role to work with the energy of the heavenly bodies. Both the inner and outer sun were honored, the inner and outer moon were honored. Through honoring both the microcosm and the macrocosm, renewal followed. Through the vision of the earth in relationship to the heavenly bodies, decisions were made as to where to build structures. A continuous circuitry of the earth and the heavens was respected and maintained.

Consecration of the Earth

We are at an interesting point in humanity's evolutionary process. We have a choice to make, and that choice has to do with our ability to feel the energy of the earth as akin to energy within our own being. Also and perhaps most importantly, it has to do with the ability to sense our own power, both our inner power and our outer power. That power can create goodness within our own lives and goodness for the well-being of the earth.

The only way we can make a change and move from what is currently happening to the earth is to participate in what is being echoed from within our souls and within the radiant core of the earth. The echo is the same. It says:

> Yes, you can live a life of goodness, you can live with harmony, for this is within your power. The echo is heard in the streams cascading down the mountains, live as you are meant to, do not search for what will not

fulfill you, rather live in accordance with the higher principles. This is what is meant for you, and this is indeed consecration.

We can hear and take heed of this voice issuing forth from all of nature. It is the same voice issuing forth from deep within our own beings, from within our own hearts' yearning. Yes, this voice says, there is no lack. There is only abundance and yet there are simple and good laws to follow. Heed these laws as if they are the foundation of life, and take care of the mother as if she is in your charge.

Desecration turns into consecration within a split instant. It is a necessity for our survival to move into that alignment.

Consecration is active acknowledgement that the planet, the stars, the galaxies, the earth, and you are not separate, but are a harmonious, luminous body of light. Consecration becomes possible when love is seen not only as something to grab at for fear that it may be gone, but rather as the vehicle in which to travel through life. Consecration is made possible when we see that we are bodies of light. As we learn consecration for ourselves, we cannot help but learn consecration for the earth.

Consecration stretches beyond the earth to the higher realms of light, to the higher spheres of light, to total love and completion. As we approach more love and light, we can issue forth this love to beings on this earth and to our mother earth. It is through our love that we can heal ourselves and heal the earth.

Our love without form imparts itself into form and enlivens the form. We heal the earth by healing ourselves. We heal ourselves by consecrating ourselves to our own holiness, and to love that is total beauty.

From light springs forth light, from life springs forth life.

Celebrations

The word "ceremony" is derived from the Latin word "sacer" which means "sacred." An aspect of the ceremonial function of the ancient structures as well as other energy centers was the aspect of celebration of life. The energy centers provided a place where the people could come together in a spirit of renewal. This joy of celebration, of bringing people together, was an important one. Through the spirit of celebration issued forth renewal. This renewal was and is of the earth and her fertility; it is of peoples and their personal patterns of life. Just as it is very possible that the earth responds to alignment through the ancient structures, so it is possible that she responds to our contemporary spirit of life and of renewal.

There exists a correspondence to the earth and to people's happiness and unhappiness, thoughts and deeds. Vibrations echo through the planets, through the galaxies during times of celebration, of bringing together. In acknowledgement that life brings forth abundance, the creative spirit of life responds. Through celebrations, the spirit of renewal issues forth.

Vision of the Heavens

There is a mystery of life to be perceived when you take time to be silent and to take in the magnificence of the stars. The following is an excerpt from the book *Guardian of the Tall Stones*, of the writer's experience of a vision of the heavens.

"In this state she was no longer aware of the darkness as darkness, but as the night sky, immensely vast and filled with countless stars. When she had looked at the sky at night on other occasions she had seen the myriad sparks of light dotted about, apparently at random. Now she was aware of it as an intricate but definite pattern.

"She saw each star linked with the others in a relationship that was unmistakable. It was as though fine gold lines, as fine as a spider's web, were drawn between each spot of light to make an exquisite network, complex, yet ultimately simple.

"But even as she grasped this, the vision was altering slightly. The web was not flat but had depth as well. The stars she had thought were all the same distance from her appeared now to vary, some nearer, some farther away. The golden threads linked not only sideways but backwards and forwards as well.

"She felt herself moving nearer to them, somehow being among them so that the network of fine gold lines was around her in every direction...stars were around her in every direction.

"As the sensation of movement grew she realized that it was not only herself moving. The stars, the golden lines, the darkness...everything was moving and everything was changing in relationship to everything, in subtle ways, at every moment. Yet the overall web of relationship was still there...the threads never broke...only adjusted, stretched and altered."*

Meditation on the Sun, Moon, and the Stars

This meditation is meant to help you consciously align with the life-giving qualities of the sun, the moon, and the stars.

With a quiet mind, allow yourself to relax and feel a sense of well-being. Picture the sun in your mind's eye, and feel the presence of the light of the sun. Think of the cycle of the sun, the rising at dawn, the early morning, the midday, the late

---

* Moyra Caldecott. "Guardians of the Tall Stones." *The Sacred Stones Trilogy.* Celestial Arts, Berkeley, CA. 1986.

afternoon, the setting sun. Feel how you are personally affected by the day's cycle. Feel yourself being energized and renewed by the sun. And feel the peace that the end of the day brings at sunset. Remember a beautiful sunset that you have seen and experience the feelings that this evoked within you. Give thanks to the sun for its life-giving qualities and for the light that it imparts to all of creation.

Now think of the moon, and visualize its different cycles: the shape and feeling of the quarter moon, the beauty and majesty of the full moon. If you are a woman, think of how your own body is in cycle with the moon to give you a time to reflect and to renew yourself. The oceans' underground currents and even your own blood move with the movement of the moon. Visualize the cooling rays of the moon, reflecting the sun's light and bringing peace to your mind. Give thanks to the moon for the deep beauty that it brings into your life.

Think of the stars, shining in the night sky. When you are able to, spend time looking at the stars on still nights. As you gaze at the light of the stars, see how that affects your consciousness. Feel what resonates within your own being when you look at the light of the stars. Imagine a lattice of energy, not only encircling the earth in a grid pattern but interconnecting the stars and planets of this galaxy and of others. Visualize this lattice made up of shimmering light, and see that the universe comes from one source and that there exists a communion of all the galaxies.

There is a sound vibration that emits from each planet, each star. There is, as well, one sound vibration that emits from the entire universe. This is the sacred sound "Om."

In ancient times different star configurations were recognized as profoundly affecting the lives and consciousness of the people of earth. One example of this configuration is the alignments that were made by the Incas concerning the Pleiades. It is possible that different star configurations are in

correspondence to our own consciousness and to our patterns of life on earth.

These alignments can also be likened to planes of consciousness that can be reached from within.

Meditative Exercise — Alignment with the Earth and the Heavens

Find a place in nature where you are free to walk and to explore. As you walk, let yourself feel the magnetic energy of the earth. See if you can feel the different energy currents from the earth. As you walk allow yourself to be guided to a place where you feel like stopping, a special energy place for you. Stand in this spot and feel the earth's energy under your feet. At the same time feel the currents of life issuing down from the skies. Your body receives both energy from the earth and from the skies. If you feel like raising your arms, palms up, to better receive the sky energy, then do so. Let your hands move as they want to so that your body and psyche can be balanced and recharged by both the heavens and the earth.

Try this exercise in the daytime with the sun out and receive the life-giving energy of the sun.

Then try this same exercise in the night when both the moon and the stars are out.

Let yourself receive vital energy from the skies and from the earth.

CHAPTER TWO

◇

The Spirit of the Earth

Gaea

"I shall sing of Gaea
universal mother
firmly founded
the oldest of divinities"
— *Homeric poem*

In Hesiod's *Theogony*, Gaea created all of creation. She is the earth mother, to whom we all belong.

"But Gaea's first born was one
who matched her every dimension,
Ouranos, the starry sky,
to cover her all over,
to be an unshakable standing place
for the blessed immortals.
Then she brought forth the tall hills,
those wild haunts that are beloved
by the goddess Nymphs who
live in the hills and in their forests.

Without any sweet act of love she produced
the barren sea."*

Gaea gave birth to the first race of gods, who then gave birth to human beings. The temple of Delphi was first dedicated to Gaea. The site was founded because intoxicating vapors that gave prophetic powers issued forth from an opening in the earth. Eventually the temple was dedicated to Apollo. Gaea has been associated with prophetic powers.

When Erichthonios established the Athenian cult, he proclaimed that every other sacrifice must be preceded by an offering to Gaea in thankfulness for her nurturing.

---

* Hesiod. Translated by Richard Latimore, University of Michigan Press. Ann Arbor.

Lessons from Nature

As you listen to nature, she will teach you.
She will show you that everything is connected and interdependent.
She will show you that life moves in cycles and that you cannot rush something coming to you until it is ready, until you are ready.
She will show you that from nature issues a sound of life, a sound of joy and thanksgiving.
This she will teach you.
Many other things she has to tell you.
Are you willing to listen?
Michael Roads, in his book *Talking with Nature*, writes about his inner communication with nature, about how he taps into the inner spirit of nature to receive nature's teaching. The following is an excerpt from his book:

> "Once triggered, my response to Nature became a daily adventure, and I spent hours by the river writing the events and words in my journal as each moment unfolded.
>
> "Scrabbling down the steep bank to the river, I felt the power-immense waiting for me. For a moment I felt inadequate, a sense of the old, familiar fear, then it was gone, lost in the surging consciousness filling all life around me. Even before I reached my place on the rock, the words were moving clearly and with undeniable force through my mind.
>
> " 'As you sit on the bank by the river, be aware of the life around you. The Kingfisher flashing color in the morning sun. Fish jumping in the clean-washed river. The sun warm on your shoulders. The water Dragon watching from a rotting log, distrust in its eyes.
>
> Microbes teeming in the rotting log...

" 'Be aware of the life beyond that which you are. Each leaf bursts with the hidden energy of spring. The force of life makes its presence known in all the creatures around you. Life in life. Life upon life.

" 'You listen to the silent song of my voice. I have spoken to mankind from the beginning of your time. In a cascade of sound I call to you from every waterfall. With loving embrace I hold your bodies in the water of the earth and softly I whisper.

" 'In endless ways I call you, but so seldom am I heard. Not with your ears may I be heard, but in your heart, your consciousness. So easily are you lost in the labyrinths of your dimension. Your minds seek always to compare. Light compares with dark, strong with weak. Always the polarity of opposites reveals the identity of your experience.

" 'I have no opposite. You cannot listen to my inner voice and then compare. You may not gaze into my inner world and seek the comfort of opposites. Your reality must stretch. Your limits must be pushed back. The dimension of opposites is but one facet of the whole reality.

" 'The five physical senses of humanity are both your freedom of expression and the walls of your prison. This need not be. Humanity has the ability to create. Creation is an expression of the power of visualization. Within this controlled, creative framework, you may open doors into the realms of Nature where I will meet you.

' "You need only open the door, I will be there. Alas, to open this door defies all but the few. There are those who step beyond the known boundaries, but you know them not. For the most part they have a wisdom which embraces silence. They sip from the nectar of more subtle kingdoms, yet they sip sadly. People such as these wish to share the nectar of their lives, but it has long been rejected. This time is coming to an end. Once again, humanity stands on the threshold of Nature's secret kingdom. The Kingdom of your own Being. A secret wide open to all mankind, yet hidden behind the veils of love, of wisdom, of integrity.' "

Nature Spirits and the Devic Kingdom

When you take a walk in the woods and you feel a lightness and happiness, perhaps it is the presence of nature spirits whom you feel. Each tree, plant and flower has nature spirits with them, tending to them, and interweaving life with them. Each, also, has a Deva associated with it. The Deva, or Shining One, is the spirit of intelligence of the plant or tree. The Devas are in charge of the form of the plant or tree. The nature spirits distribute this intelligence to and throughout the plant or tree. You can communicate with both the nature spirits and the Devas when you extend your friendship to the plants, trees and flowers. The Devas assisted those in Findhorn in the development of their garden. Even if you cannot visually see these subtle life forms perhaps you can perceive them with your subtle eye and heart. Nature spirits and the Devic Kingdom are happy because they exist in cooperation with life. Rather than trying to block things and to make them over, they celebrate life as it is in its true splendor. The nature spirits and the Devas want to communicate with human beings so that humans understand more of the presence of life, and how to simply *be* rather than to control or dominate. They want

---

* Michael Roads. *Talking with Nature*. H.J. Kramer, Inc. Tiburon, CA. 1987.

people to know that all life comes from one source, and that happiness lies in being in harmony with the source of creation.

So, when you are walking in the woods, or when you are growing a garden, use your senses to perceive life that is all around you.

Presence of the Angels

In certain places in nature there are angelic beings who can be seen with the subtle eye. Angelic beings are not associated with any particular form of life. Rather, their role is to assist all forms of life. The angelic beings are not bound by the laws of nature, but do exist in nature. This is an interfacing between the higher realms of light and the creative force of life. Angels can be called to assist anywhere that you live. Certain places, however, have energy currents that are more conducive to welcoming the presence of angels.

The most conducive place of all, however, is in your heart.

Revelations from Nature

Revelations from nature are a part of humankind's connection to the creative nature of the universe. All of life is linked to God's light. Revelations can come in the form of visions, dreams, or the inner voice. Nature can be thought of as a reflection of the laws of creation. Although it is not necessary to attune to nature to receive higher guidance, nonetheless it is a powerful way to join our consciousness with the spirit of all life.

In many cultures it was considered a rite of passage to have a vision quest, to spend time totally alone in nature, in a spot chosen by the elders as a sacred place of initiation or in a self-chosen spot. Often fasting occurred at the same time so that the participants could both sharpen their senses and purify their bodies. From this time of solitude, and of humility, would come a vision. This vision could occur in the form of a bird whose spirit would teach, or from an animal, a rock or the wind or water. Although the guidance came from within their own beings, often it was linked to a place or to the spirit of nature.

From this experience each emerged a new being, one who had submitted him/herself to a time of solitude and was transformed into a more powerful being. Places of power where vision could occur were known by native tribes and respected.

Many power places are also places of vision, and places of initiation.

Horticulture and Telepathy

Several books have been written fairly recently about developing horticulture telepathically. One way these pioneers work with the Devas is by telepathically communicating with the Devas.

Michael Roads, in his book *Talking with Nature*, recounts an experience he had when he was a farmer in Australia. He was having considerable trouble with wallabies eating his pasture. Finally, after attempting to poison and to shoot them, he decided to communicate with them telepathically. He said to them, "I don't know if you wallabies can hear me, but I am offering an agreement with you by which we each meet our own needs. I am asking you to stop eating our pasture, and in exchange for this I will see to it that nobody shoots you again.

However, because I realize I must share this land with you, I will allow you to graze around the outside of the paddock. Please don't take more than twenty yards." To his surprise this arrangement worked.

He goes on to say, "We recognized their divine right to life, realizing that cooperation with Nature offers unlimited potential."*

Crystals and Mining:
Interview with Gary Fleck

Gary Fleck mines crystals in Arkansas. He is a professional mining engineer who teaches about the healing properties of crystals and other stones.

"My most profound experience in mining crystals was seeing and being taught by an angel of inner earth. We had gathered together in an underground mine in Arizona to do a prayer ceremony. During the prayer ceremony an angel came to us and told us she was an angel of the inner earth. She taught us about crystals, their usage, and the evolution of the earth in general.

"She also talked about ley lines, the grid, and power centers. Crystals are our friends, they are here to teach and to heal. If you choose to use a crystal, then ask the crystal how to do it. If you truly do want to work with crystals, then you have to go inside and get in touch with your own feelings, and your own spiritual guidance.

"I don't feel that crystals should be mined for a long, long time. Hopefully as our consciousness grows we will not need to work

* Michael Roads. *Talking with Nature*. H.J. Kramer, Tiburon, CA. 1987

with crystals as much. Right now we need the three-dimensionality of crystals.

"Jesus said, 'By the power of your faith be ye healed.' We want to look beyond the crystals and look towards Christ-manifestation. The crystals are playing a role to attune to that energy at this time.

"What it takes to heal the earth and what it takes to heal ourselves is simply a willingness to do so. If we exhibit a true willingness in our consciousness, then the ways to work with it will be provided for. You can't heal the people on the earth without healing the earth itself, and you can't heal the earth without healing the people.

"As long as we take from the earth with respect, instead of raping her, then we can maintain a balance. The earth is our mother and she wants to give.

"Give thanks every day to the earth. You love the earth. You utilize the earth in the most conscious ways.

"This is perhaps the most healing thing."

Caring for the Land

Just as you would care for your own eyes, it is necessary to care for the land. Think of the earth as an extension of your own body, and care for both your body and the land.

How can you care for the land?

First, in very personal ways. Don't use more than you need. There are those who make money telling you that you need more than you actually do. When you buy things, and waste the earth's resources, you end up feeling disappointed and bitter. Do not be used by others for their financial gain, and in the long run, ruin both yourself and the beautiful earth. Use

what can be recycled when possible. Respect all that is given to you.

Second, care for the land where you are. Take an active interest in how the resources are being used. When you eat food, give thanks to the earth, to the Lord, for providing for you. If you are guided to learn horticulture and the higher laws of horticulture, do so. When you practice horticulture, listen to the spirits of the plants. They will tell you what to do.

Third, understand what mining and oil drilling does to the earth. Become aware of what is mined, how the mining takes place and who profits from the mining.

Learn what deforestation, stripping away the earth's forests, is doing. Feel how the earth cries daily, for you are part of the earth and her well-being is your responsibility. In the Amazon rain forest the creatures are losing their homes; earth is losing its most vital skin.

Fourth, send prayer to the earth, send thanksgiving to the earth. Be kind to others, for the earth, as all of the stars in the galaxies, resonates with your thoughts and your actions.

Care for the earth. The earth's wealth is our wealth. The earth's devastation is our loss and devastation. We cannot be healthy if the earth is sick.

Learning to care for the earth will raise us to the level of participants, rather than those who exploit and then discard. Listen to the birds, for they have wisdom to impart to you. Look at the stones, at the rocks, for inscribed in them are teachings: Remember God who created the earth and observe in thoughts and in action the holiness of His creation.

Go to sacred places and pray for healing and understanding, but remember that wherever you are is sacred.

The Spirit of the Forests: The Sacred Tree

The tree has been the symbol of the foundation of the universe in many many cultures. The Old Testament tells of the tree of knowledge of good and evil whose fruit drove Adam and Even out of the Garden of Eden. Here the tree symbolizes the movement from unity to duality. This movement is a necessary one for consciousness to perceive its own divinity. For without knowledge of the pairs of opposites, how can one know unity, that which is beyond the pairs of opposites. The medieval text of the Kabbalah, the *Zohar*, says, "Happy is the portion of Israel, in whom the Holy One, blessed be He, delights and to whom He gave the Torah of truth, the Tree of Life. Whoever takes hold of this achieves life in this world and in the world to come. Now the Tree of Life extends from above downwards, and it is the Sun which illumines all."

Isaiah wrote about the Tree of Jesse: "There shall come forth a shoot from the stump of Jesse, and a branch shall grow out of his roots...In that day the root of Jesse shall stand as an ensign to the people" (Isaiah II:1,10).

In the Christian Western tradition, the Christmas tree symbolizes renewal, spiritual birth, and hope for Christ consciousness, or the highest precepts to manifest on earth.

In the Buddhist tradition, the Bodhi tree is the tree under which Buddha sat and achieved enlightenment. This is where he overcame the delusion of duality, spoken about metaphorically in the Old Testament as the Tree of Knowledge of Good and Evil. There is an ancient legend that the Buddha's mother gave birth to the Buddha in front of a flowering sal-tree. While she was giving birth, she held a branch of the tree in her hand.

> "And I saw that the sacred hoop of my people was one of many hoops that made one circle, wide as daylight and as starlight, and in the center grew one mighty

flowering tree to shelter all the children of one mother and one father. And I saw that it was holy."*

"For the Mayans the trees are our brothers. The word for tree is Te, turned around the word is Et which means brother. K' ik is the name of resin in the tree. This is the tree's blood. K' ik is also the name of blood in human beings."
— From an interview with Hunbatz Men

Impressions of the Amazon Rain Forest: Interview with Rusty Crutcher

Rusty is a professional musician who took a trip to Machu Picchu and the Amazon in August 1986. There he recorded two cassettes entitled "Machu Picchu Impressions" and "Amazon Song." He has a company entitled "Emerald Green Sound Productions" in Santa Fe, New Mexico.

"We took a boat ride down the Amazon River to the Tambopata Reserve. I felt sad. I felt a crying out of the jungle, a choking of the jungle, the rain forests. "There was a 'slash and burn' program happening at the time where they burn the rain forests down to increase crop production. They don't realize that the topsoil is only a half-inch thick. They burn the rain forests down, the rains come and this half-inch of topsoil erodes right off like nothing. The rivers are muddy and jammed with gook. That's the soil. It's gone. It will take years and years to come back to its normal state. It creeps in from the sides and

---

* *Black Elk Speaks: Being the Life Story of a Holy Man of the Oglala Sioux as told to John Neihardt.* Lincoln: University of Nebraska Press. 1961.

has to recompose itself. It's a very slow process. When the trees are burnt down, the rain fall diminishes also. It changes the whole cycle.

"The jungle is going, the rain forest is going. You smell the burning. You'd see the birds there — the Macaws, exotic parrots. Their cry wasn't a happy one. There were scientists there, people who were concerned and who were doing studies.

"We are all affected by this.

"The rain forests provide us with 1,400 plants that assist in curing cancer.

"The tropical rain forests assist in regulation of both droughts and of floods. The specific trees of the rain forest, when they are burnt, release carbon dioxide in the air, that many scientists link to the greenhouse effect. As the greenhouse effect continues and increases, the distribution of rainfall around the world would change. Norman Meyers in his book, *The Primary Source*, says that a temperature increase of just 2% F. might reduce the U.S. corn crop by at least 11% if it is linked with a decline of August rainfall. Scientists have noted that there has been an increase of global temperatures by approximately 1% F., and that half of this increase has taken place from the mid 1960's on."

Shaman as Ecologist

The Desana Indians, a part of the Tukano Indians, live in the Northwest Amazon. The Tukano believe that their land was handed down to them by their forefathers in perpetual trust. They see their universe as having finite resources. There is a continuous circuitry between humans and animal or plant nourishment. The Tukano believe that a human should not use

energy without somehow replacing it, or restoring it. They believe that people must conform to nature and must tailor their own needs according to what is truly available in nature. The Desana is conscious of the effect his/her actions have, not only on society, but in the entire universe. When illness occurs, it is thought to be a result of actions taken that are in disharmony with tribal needs as well as with nature. It is true that where the Desana live, there is a limited supply of food and other resources. If too many animals are killed for food there can be no replacements.

The Shaman's true role therefore is as an ecological regulator. Shamans hold that overhunting of one species is one cause of disease, the depletion of plant resources another.

The Shaman actually is responsible for managing the resources of society. Depletion of the natural resources of the rain forest is considered one of the worst offenses a person can commit.

The Shaman is in communication with the spirits of the animals and the plants. He plays an active role in the construction of a communal house, in the decision of what fish to throw back into the water after a catch, and in all of the uses of the resources of the tribe.

The Desana know that maintaining ecological balance is vital to their continuation.

G. Reichel-Dolmatoff "Cosmology as Ecological Analysis: A View from the Rain Forest." Huxley Memorial Lecture. Royal Anthropological Institute. 1975.

The Body Physical: Establishing Harmony within The Body

Just as there is harmony in the celestial body, including the earth, so must there be harmony in the physical human body. Within the physical body exists the dwelling place of the spirit, within the dwelling place of the spirit exists the most secret of all secrets. The physical body therefore is to be understood as the temple of the most holy, the dwelling place of the most mysterious, the sacred flame of God

Although the body is made of perishable elements, what dwells within the body in the form of consciousness does not perish. The body therefore can be cared for as the dwelling place of the most holy and not worshipped as an end in itself. For the body is made up of the elements of the earth and will return to the earth. To care for the body is to respect the earth, and to respect the body is to care for the earth.

To establish harmony within the body is a continuous and transformative process. The most important element is one of awareness of the body and the signals that the body is giving. There are very simple laws that the physical body must follow to maintain or to regain health. The earth and nature are healing forces and, if treated with respect and awareness, can help to heal the body.

Running waters such as streams, waterfalls, and water of pure lakes all contain healing energy within them. Water such as hot springs also helps to purify and to balance the body.

There are healing clays and places of special magnetism in the earth that have strong healing properties. Clay from Arkansas that forms around quartz crystals has certain properties that help to draw out and to recharge. Since ancient days physicians have known and used the healing properties of clay. The great physician Avicenna used clay for his healing work. The Greek, Galen, also used clay for healing purposes. Pliny the Elder of

Rome wrote a chapter concerning clay in his book, *Natural History*. In more recent times naturopaths such as Kneipp combined the use of clay in their treatments.

A friend recounted her experience with healing earth: "Several years ago I went to New Mexico and developed an infection on my skin that just wouldn't get better. At Chimayo, New Mexico there is a small sanctuary dedicated to the Mother Mary. In the sanctuary there is a hole of healing earth. I placed some of that earth on my infection and within 2 hours I watched the infection healing itself."

Give the Earth Nourishment:
Interview with Don Juan Camargo Huaman

Don Juan Camargo Huaman was born in Cuzco, Peru, the ancient capital of the Inca Empire. He mastered the seven stages of Initiation of the Inca tradition and has devoted his life to spreading the wisdom of these teachings under the ordination of the Great White Brotherhood.

"The earth since old times is thought to be a living being. She has an astral body, a mental body, a spiritual body. She is also a being in the process of evolving. The earth can't evolve when she is sick. It stops her growth. Her sickness also makes humans sick. That is why people are so crazy now.

"The earth is like a mother who needs the affection from her child. The earth doesn't protect herself. Like a mother she just keeps on giving. In this way, her children should stop abusing her.

"Our tradition shows salutation to the earth. Through our prayer, our dance, our song, we show her our love. We must take care of the earth better. She is getting worse. A great part of humanity is sleeping. They do not realize how terrible it is. In all the major countries science knows the destruction that is going on, but they don't stop.

"Give the earth offerings, give her strength that she needs. She needs our energy, our love, for us to be closer to the earth. In ancient times, when people did agriculture, the best and the first fruits were given back to the earth. Now machines do it. Before, they begged the earth, they thanked the earth, there was ritual done, they sang, they danced. Agriculture was done spiritually. Man was respectful and thankful. Now, man does not even know how to take a plant. There are guardian spirits of the earth who protect the earth. When there is too much abuse, calamities happen.

"If you pray to the earth and you are full of pride and full of vanity it does not good. Purify yourself from within. People talk about peace, about love, but many of these people cannot experience a moment's peace. Spirituality is not just talk; it is your actions, it is the purity in your heart.

"Machu Picchu was a spiritual school. All of the temples were connected to the sun, moon, and constellations, according to their work. There was an astral relationship, an anatomical and a spiritual one.

"Even the people who lived in the city did not know about Machu Picchu. The initiates in Machu Picchu knew the Spanish were coming beforehand. They hid everything about 500 years before the Spanish came. There are actually bigger and greater ruins that Machu Picchu that have not been discovered yet. There are greater places than Machu Picchu where people still live and study that are not known about.

"Crystals are most precious to the earth. They are similar to the endocrine system, the glandular system in our body. People should only use crystals for good, not just for decoration."

A Ramage for Waking the Hermit

"Early in the morning the hermit wakes,
hearing the roots of the fir trees stir beneath his floor.
Someone is there. That strength buried
in earth carries up the summer world.
When a man loves a woman, he nourishes her.
Dancers strew the lawn with the light of their feet.
When a woman loves the earth, she nourishes it.
Earth nourishes what no one can see."
— *Robert Bly*

*Reprinted with permission of Robert Bly. 1985.

The Five Elements

The five elements, earth, air, water, fire and ether make up our physical universe. Each element supports the other, nourishes the other, and helps to form a basis of the physical universe, of the physical vehicle of spirit.

Take time to know these elements, to appreciate these elements, to pay homage to these elements. Your physical existence depends on these elements.

EARTH is the source of our food, she is the one who nourishes us through her giving forth abundantly. Her trees provide us with air and rain; without the forests we cannot remain alive. Her vegetation provides us food, her fertility brings forth plants to heal our physical diseases, her flowers give us joy and well-being. The earth reveals her mysteries to those who can listen. Rocks provide the earth and its inhabitants with endurance. The rocks of the earth provide us with inner listening, with spiritual attunements, with strength. The minerals of the earth are the earth's skeletal system. The crystals are the endocrine system. The oil is her precious fluid that gives her vitality.

The green of the earth provides nourishment to our very souls.

WATER is our life's source. Our bodies are made up of approximately 50% water. Through water, we are able to live. Through rain, food is grown. Regulation of rainfall is provided through the forests. Without the forests and the rain we could not live. The streams of the earth are her circulation; when they meet into rivers there is rejoicing. The rivers of the earth are also her circulation; when they join the ocean there is the joining of the small into the great, the limited into the vast. Lakes in the earth provide us with inspiration, and link us to our higher selves. The oceans show us our unlimited nature. The oceans provide us oxygen.

AIR links us to the spirit of God. As we breathe in air, we revitalize our cells; we give our bodies and our psyches new life from this element. As we exhale air we release toxins that have accumulated and the air itself purifies our systems. Without breath there is no life. The wind sends us messages. Our soul's yearnings echo in the wind. As we listen to the sound of wind we hear our own inner yearnings, our own heart's voice. The wind circulates life's flow. It is not idle movement, rather it is movement that nourishes the earth, that nourishes the water. The wind welcomes us home to ourselves.

FIRE is the knowledge from which life springs. The spirit of God is fire; life forms can be consumed in fire and yet the fire retains its essence. Fire teaches what is behind forms, what is eternal. Fire teaches us about ourselves. It is the element that purifies, that changes base metal into gold. Since very ancient times, yagnas, or very special fire ceremonies, were performed to give blessings to the people and to the earth. Offerings were made to the fire, and the fire received these offerings. When fire is not treated with reverence, it can cause destruction. Respect the fire, and see the work of the creative spirit of God within the fire.

ETHER is the support of the earth, of the heavenly bodies, of that which manifests on earth. It is the unseen element behind the seen. It is space. Within ether, all other elements exist. Without ether, nothing exists.

Meditation —
The Earth as a Body of Light

Visualize the earth as a body of light. She is surrounded and permeated by the golden light of God that renews her, recharges her, and maintains her as she makes her shift into her new place in the galactic field.

Love is the energy of sustenance; it is the foundation of all being and all creation.

Think of the earth with love. Send her thoughts of love and of blessings. Visualize her as a planetary body of light.

CHAPTER THREE

The Grid of the Earth: Sacred Sites

Healing with the Earth: Giving and Receiving Energy

The earth has the capacity to heal us and we also have the capacity to heal the earth. Here once again the Law of Reciprocity is applied, and through this a continuous life current can be cultivated. Although this circuitry is a continuous one, it is easy to begin by doing a simple visualization exercise. If you can, it would be good to be on the earth, not wearing shoes when you do it.

Visualize your body bringing down cosmic life energy from the stars and galaxies through the crown of your head. Send this energy, the color of white light, through your body, energizing your own body, and then through your feet into the earth. As you do this send love to the earth.

As you stand on the earth bring the electro magnetic energy up from the earth through your body to revitalize and heal it. Visualize this energy balancing your body as you walk and as you stand and actually strengthening its circuitry. If you are in the woods, see the life that the earth sustains such as the green of the trees permeating your subtle energy field, and bringing more life and circulation to yourself.

Magnetic Energy Flow in our Lives

Just as the earth has energy currents that need to flow uninterrupted to create health and vitality, so our own lives have energy currents that flow in a way which creates health. As we learn intuitively to sense the energy currents that flow in the earth, we can also learn to sense how energy flows in living our lives. Instead of resisting energy and trying to dominate the flow of energy in our lives, we can see how the energy really wants to move, what direction it really is taking and go with that flow.

When we are able to do this, our lives work out in a natural way as part of the divine plan.

Energy Currents in our Bodies

Just as energy currents in the earth must flow to maintain health, our bodies have energy currents, called meridians, which must flow to maintain health. While blockages in the earth's energy currents are often caused by pollution, mining, drilling of oil, human blockages are caused by thoughts and feelings, as well as the toxicity of the environment.

The mind can be called up to create flow within our bodies and our psyches. We can pray to God and visualize the cells in our body regenerating and becoming healthy. As we attune more to subtle energy we can perceive how to encourage flow and vitality. As we attune more to subtle energy we become more aware of the earth's need for our help, for our consciousness and effort to help her regain her flow of life.

How does this healing work? The human body has subtle energy currents that connect the etheric to the physical, so also the earth has energy currents that act as pathways to the earth's cohesiveness.

The Earth's Energy Currents

Right beneath the surface of the earth exist energy currents that criss cross, interweave and circulate. These currents are made of more subtle matter and look like shining threads. These energy currents which have been felt by dowsers are a kind of nervous system of the earth's energy body. When there is an interruption of these energy currents, as in the case of mining or waste disposal stations for nuclear power plants, the energy of the earth is interrupted. When this occurs the life force of the earth goes into a kind of stagnation.

The Celts Goddess, Elen Helena, is the oldest and first of the Celtic deities. She is considered the Goddess of the Ancient Tracks. From the natural energy currents, roads developed in ancient times. These roads were called "old straight tracks."

The Grid Pattern

"All key lines lead to the planetary Grid, the primary light and energy matrix, creating, enveloping, and maintaining planet earth, our Gala. The Grid has been variously described by poets and clairvoyants in recent times. A personal acquaintance of mine in England summed it up lucidly: 'I saw the earth as a fishnet web of light lines. My body was the same, and there were lines of light radiating from the intersections of the planet's surface to link with the web network surrounding other planets.'

"This is because the primary message the reality of the Grid is telling us is this: We are all one crystal-line vibrational body — galaxy, solar system, earth, human."*

---

* Richard Leviton. *Ley Lines and the Meaning of Adam.* "*Anti Gravity and the World Grid.*" Adventures Unlimited Press. 1987. Box 22, Stelle, IL 60919.

The earth has a matrix of energy patterns that sustains her, orients her and holds her. This energy matrix is called the Grid of the Earth. There is a crystalline grid within the earth and there is also a subtle etheric grid around the earth that interpenetrates the earth to help the earth vibrate as a cohesive whole. The grid around the earth can be thought of as the mental body of the earth. It is a manifestation of the consciousness of the earth. The higher octaves of light interface through this energy grid. Certain points on the earth that are interpenetrated by this grid are known as power centers or as sacred places. The ancients intuitively understood about the earth's grid and built their temple and places of worship on points of the grid.

Tom Graves in his book, *Needles of Stone*, suggests that the structures on some of these sites, such as Stonehenge, functioned as a kind of acupuncture needle. Buildings or the stones themselves connected the cosmic energy from the heavens and stimulated the earth with this energy.

Understanding and attuning to the grid is necessary for us for a few reasons. First, if we understand that there is indeed an interpenetrating network of subtle energy around the earth, we also understand that we are, in fact, connected to the realms of light, the higher octaves of light, and to the earth as a whole. We feel that what happens in the Amazon Rain Forest, where trees are so brutally torn down and burned at a devastating rate, will directly affect us because everything on this earth is interrelated.

To attune to these special places of light and power is a very personal quest. If you take time to be still, you will intuitively feel what energy centers on the earth you are drawn to. Or someone will come along in your life and act as a catalyst for you to discover a sacred place. Energy centers that are mentioned in this book are but a few.

Earth crystalline energy grid

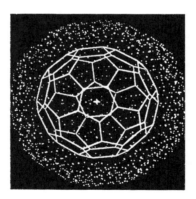

Note similarities in grid concepts with this computer model of a carbon molecule.

*David Hatcher Childress *Mapping the World Grid.*
*Anti-Gravity and the World Grid. Edited by David Hatcher Childress, Adventures Unlimited Press. Box 22, Stelle, IL.

> Listen to the Earth
> And She will teach you.
> Listen to your Heart
> And you will be guided.

We also can intuitively feel the major significance of the grid pattern because of the shift that is occurring in the earth's vibrational field. This shift of the earth not only affects the earth but also directly affects our consciousness, for it is a time of great opportunity for spiritual growth.

At certain points on the grid pattern there are energy centers that radiate light and act as energy vortexes. What is happening now is that the energy points on the grid pattern are shifting and so some energy centers are more important than they were previously. This reactivation of the energy centers creates more balance for the earth and assists in bringing down cosmic energy that activates the energy field of the earth. Not only do the energy centers activate the magnetic field of the earth by interfacing with higher energy patterns, the energy centers also activate our own process and understanding.

Some energy centers that attune to the higher octaves of light are places to receive guidance and revelation from within. Others are sites of ancient structures and point us in some evocative way to an understanding of the connection of the earth and the heavens.

Still other energy centers provide us a place to recharge and to heal. They are more magnetic in nature. Energy centers can serve people who visit them as a way to attune to their own higher fields of light. If there is prayer sent to the earth with a pure heart at these centers, the earth herself will be strengthened.

The blueprint for the earth's evolution lies in the crystalline-mineral-ore grid within the earth. This grid is stimulated and changed by stellar events. This blueprint has been in the earth and is the foundation of the earth's consciousness.

This is one reason why people are attracted to going on pilgrimages to sacred places.

Meditation

Visualize blue light forming a grid pattern within the earth. The function of this grid pattern is to keep the earth strong. See the special centers of both ancient times as well as centers that are becoming apparent now as points on this grid pattern where information is both received and recorded. With your heart, send the earth light through this grid pattern.

Crystalline Grid and the Australian Aboriginal:
Interview with Lorraine Mathi Williams

Lorraine Mathi Williams is one of the hundreds of Australian Aboriginal people who has specialized training in the earth, the grid, the environment, the heavens and crystals. She is also a film maker, director, actress and story teller. She is an advisor to the Education Department on the Aboriginal culture in Australia.

"We refer to our origin as dreamtime. Prior to 3 million years ago, our people say that we lived on another planet that suffered a planetary disaster. Some of the refugees went to live on other planets; seven spirit brothers from that planet came to earth and this is where we derived our spiritual ancestral being. That spirit being's name is Wandgina.

"The descendants of the seven spirit brothers each had a responsibility. The Australian Aborigines' responsibility was to the earth...to erect ley line grids, energy grids. They were taught by helpers from other planets.

"There are still helpers from other planets on earth. There are 1,000 representatives from nine planets today helping out in all sorts of fields.

"The rainbow serpent is made up of all the minerals in the earth, the gold, the silver, the copper, the rubies, the diamonds. The colors of these minerals give off an aura which is reflected up in the sky as the heart circle.

"Over the years, over the thousands of years, man has broken down the energy grid and the earth has no protection. The earth goes through a galactic force when the planets line up. The earth gets buffed around and needs the crystalline grid system to keep itself together. The planetary helpers kept this energy grid activated with solar energy and lunar energy. The only things that can bring the energy from the heavens to the earth are crystals that are in the earth. Also the crystals are the cosmic galactic energies rolled into one. It is the only form that the representatives from other planets can travel on. It is food to them, it is their fuel. It is their space ship, I suppose you can say. The beings are so highly advanced that they can materialize as human beings.

"That is why I say to leave the crystals and the minerals in the earth. I also say to place the crystals in the earth. We've had all the mining done in our land. The last thing to get mined that would destroy the grid completely is uranium. Every 7,000 years the earth goes through a shattering earth change. That's the time when the planet earth is being buffeted between the galaxy and the planetary atmosphere. The decision by the planetary helpers is to save the earth and to help it through the next big shift that has already begun. We are at the beginning of a 7,000-years shift. The axis has shifted a bit, but we've got to go through one more shift.

"It was after the second big earth shift that the continents formed. When the early people erected crystal grids they had to have huge activators. Four mountains on the four corners of the earth house these activators. Our country, Australia, has one such mountain, Wullumbum. I don't know the other three, but I want to find out. Oddly enough, when the earth was one big land mass, Wullumbum belonged to the Hopi tribe. An old lady from the Hopi tribe sent a white woman to find the lost holy mountain. She found me and I took her to the mountain.

"Ayers Rock gives out spiritual energy. Huge chunks of rock like Ayers Rock are for pure spiritual energy.

"The other force to bring the energy from the heavens to the earth and to revitalize the grid is women. But men, too, must help with the grid system. The spirits have moved people all around the world to converge on certain sacred sites. You see, we human beings are the link between the heavens and the earth.

"According to the Aboriginal prophecy for the future, as soon as all of the nations come together and the circle of elders start to teach the ancient laws, the energy grid will also be complete."

Sacred Sites

"Sacred sights we can see as being at key nodes in the energy-matrix; when the structures at those sites are used for true worship, for reinforcement of and respect for our relationship with nature, they direct the energies passing through those sites towards creating and maintaining the harmony and health of the area."*

There are many sacred sites on the earth. Each site is linked to the other through the grid pattern. Sacred sites such as Stonehenge, Avebury, Glastonbury, Newgrange, the pyramids in the Yucatan, Newgrange, Machu Picchu, are sites that have structures on them. Other sacred sites, such as the holy mountains, Sedona Arizona, Ayers Rock in Australia, Four Corners in the United States (where there is uranium), Haleaka Crater in Maui, have no structures. The structure is in the formation of the site itself.

There are other special places that give our healing energy. These may be places that have a lot of mineral deposits under the surface, such as Bisbee, Arizona, where copper minerals such as Azurite, chrysocolla and turquoise are found, Hot Springs, Arkansas, where crystal deposits and healing springs are found, or Ashville, North Carolina, where a very high concentration of minerals, such as Beryl, is found.

---

* Tom Graves. *Needles of Stone Revisited.* Gothic Image Publications. Glastonbury, Somerset, England. 1986.

Approaching Sacred Sites

To go to a sacred site is to go on a pilgrimage. There is inner preparation. Contemplate why you are drawn to making a trip, and look within your heart to see what you are seeking. With a quiet mind and a pure heart you can hear the teachings of all wisdom.

When you go to a sacred site do not have expectations of what you need to receive there. Rather, spend time at these energy vortexes just to be still and to listen. If you go with a group, try to establish and maintain harmony within the group.

At most sacred sites there are spirits who watch over the site. If you bring an offering to the site, such as sweet fruit or flowers, and pray with your heart to be received there, the gates of welcome may open to you. If at any time you get an uneasy feeling that you are not welcome there, it is best to leave with thanks.

After you spend time in prayer or in personal or group meditation, remember to send your prayers and blessings to the earth and to humanity.

Holy Mountains

Mountains have been conceived of as being holy since very ancient times. Old Sumerian astronomical observers conceived of the universe in the form of a majestic mountain coming out of a sea that knew no boundaries. The structures of the Sumerians, the "zaggurats," rose up like the mountain and were considered to be as lofty as the mountain. The graded stages in the "zaggurat" corresponded to both human consciousness and to the planets in our solar system.

Mt. Sinai in the Old Testament is where Moses witnessed the eternal fire and received the Ten Commandments. In the New Testament, the transfiguration of Christ, wherein Jesus manifested in blinding light to his Apostles, took place in Mt. Tabor. Mt. Athos in the Greek Orthodox Church is considered to be a holy mountain where many visions have been perceived. The Taos Mountain in New Mexico is the holy mountain of the Indians who live there, and is protected by the Indians.

Mt. Kailasa in the Himalayas is regarded as the abode of Shiva, the primordial Lord.

Mt. Kailasa

The word Mt. Kailasa means "crystal shining." Mt. Kailasa has a configuration of a stairway that looks as if it goes right up to heaven on the south side of the mountain. The shape of this configuration is also pyramidal in form. The Tibetans as well as the Indians worship Mt. Kailasa. In Tibetan language the word for Kailasa is "Tisa" which means ladder. The Tibetan deity who lives on Kailasa is known as Demchok. His attire consists of tiger skins and a garland of skulls. Shiva is known for similar attire.

The scripture the *Mahairvani Tantra* says, "Shiva dwells upon Kailasa but one finds him wherever his worshippers are."

The *Ramayana* says, "Beyond the Himalayas is an open space of approximately 400 miles without high mountains, rivers and trees and living beings. Further on is Kailasa Mountain, the lovely quarters of Kuvera, made by the celestial architect, Vishwakarma, where is spread out a great lake full of lotuses and lilies, with swans, ducks, and heavenly nymphs playing about. Further north are inaccessible caves of great and

illustrious sages, effulgent as the sun, adored by the Gods, whose forms they take."

The great saint Shankaracharya (8th century A.D.) received the divine revelation of the Saudarya Lahiri at Mt. Kailasa. Manasarova Lake, the lake adjacent to Mt. Kailasa, is also revered as a holy lake. It is here that the ashes of the great being Mahatma Gandhi were immersed after his death in 1948.

Sacred Sites of the Mayan:
Interview with Hunbatz Men

Hunbatz Men was born in the Mayan community of Uenkal, Yucatan, Mexico. He organized the walk from Mexico to Peru in 1982, the first in modern times. Also, he organized the Cuahtemoc Spiritual March, from Mexico City to Los Angeles. Both were for the purpose of cultural unification. He lectures and gives workshops on his culture across the United States, Mexico, Central and South America. He is the author of the book *Mayan Science and Religion*.

"The Mayans are very ancient people. In our tradition, we speak about coming from Atlantiha, which is called Atlantis in English and Atlantida in Spanish.

"We have to learn to attune to the Mother Earth. She is going to advise us how she will be changing. She makes a lot of different sounds, and we have to understand the different sounds that she makes. The sounds are vibrations and we have to learn how to catch the vibration. Through meditation we can learn how to listen to and work with Mother Earth.

"We need to change when the Mother Earth changes. She will always advise us about the transformation that she is making. You can say to the mother, 'I'm going to try to help you. Please

help me to understand more about you because I am your son, I am your daughter.' We have in our real memory, memory that exists deep within us, what has happened to the Mother Earth thousands and thousands of years ago, and also what is going to be happening with her. We are going to wake up that information and to the vibration of the Mother Earth. Only in that way we will not be afraid when the big changes come. The Mother Earth sends energy through the principle points of the four directions. You can feel where the energy is coming from. You feel the energy right at the point at your navel. You feel the energy of the message there. In Mexico some traditional people know when earthquakes are going to come. There are points outside of Mexico City that indicate where and when an earthquake will come. There are lakes there that change color and temperature a few days before an earthquake. In Los Angeles there are also points that will advise people about earthquakes.

"Perhaps you remember there was a big volcano eruption in Mexico a few years ago. First the animals left that area, then the Indians left that area. People who stayed were killed from the volcano.

"The Mother Earth has memory too. Everything that has happened to her and everything we do to her is recorded in her memory. The connection of the memory of the Mother Earth and our understanding is made through the food we eat — the potatoes, the tomatoes, the mangos. Everything that we eat comes from the Mother Earth. If we don't eat, we die. The respiration of the Mother Earth is the wind.

"A new cycle of Mother Earth, synchronizing time and the human being, has begun. For this reason, the pyramids of the Yucatan are giving forth the information impressed in their millennial stones. This teaching can be comprehended by any human being. One has only to prostrate oneself in front of the pyramid and the information will begin to flow. Our cultured

Mayan people designed it so that the temples of initiation would guard the ageless knowledge. Our Mayan teachers knew that with the passage of time people would return to this land of Mayan, to regain the knowledge of these stones. They are vibrating with the new era, to teach that which was written there by the great Mayan initiates to aid humanity. For example, Uxmal is the place where women can learn to work with the energy of the moon and the Mother Earth. Uxmal means eternal moon. The planet of the woman is also the Mother Earth, so the women will teach us about the Mother Earth.

"The pyramid of Chichen Itza shows us seven triangles of light along the stairways at the equinoxes. This symbolizes our seven energy centers, or chakras, of the body. It also symbolizes seven solar systems that the earth's solar system travels with. Scientists only know of one solar system, but there are seven. Each energy center in the human being has an equivalent of one solar system. The name for the center of the seven solar systems is called Tze K'eb, the tail of the serpent. The seven solar systems revolve around the Tze K'eb every 26,000 years. The seven solar systems also revolve around the Milky Way. The symbol for the center of the Milky Way is GE. This symbol is found in many of the Mayan inscriptions, pyramids, and codices. The seven solar systems, including the earth's, have only revolved around the center of the Milky Way, GE, 20 times. The Mayans constructed the locations of the sacred places according to the movement of the stars through time. Different sacred sites have different vibrational frequencies at different times. The sacred places will tell us when we need to be there. Perhaps 2,000 years ago a particular sacred site was used, and then due to the movement of the stars, the site will be activated once again.

"If we send up armies in the sky we are offending the cosmos and the Mother Earth. We need to be clean in everything we

do. We need to be in harmony with the cosmic systems we are part of. People are going to have to change because as we move around the seven solar systems, the energy is different and it will change us."

A Prayer to the Earth

In your heart see the earth in her place in the solar system, in our galaxy, in the universe. See the earth as a living being who is evolving. She is one who gives nourishment. Pray for her to heal and strengthen herself so that she may continue to live, and to serve her purpose.

The oceans, who give us oxygen, the beautiful forests of the earth, the protective layer of the ozone, the mountains and the flat plains all act in harmony for the life cycle. Send your good wishes and blessings to all of these beautiful aspects of life on our earth.

Send your love to the creatures who live upon the earth and who partake of her nourishment and of the life cycle.

Link your heart to the men, women and children who live on this earth.

Hold the earth within your heart and pray for her renewal.

PHOTOGRAPHIC ACKNOWLEDGEMENTS

Front Cover. Valley of the Gods. Southern Utah. Bill Davis.

Preface. Valley of the Gods. Southern Utah. Bill Davis.

Psalm 24. Clouds. Taos, New Mexico. Bill Davis.

CHAPTER ONE. Morning Clouds. Taos Valley. New Mexico. Bill Davis.

p. 3. Cedar Mesa Indian. Utah. Bill Davis.

p. 6. Pueblo Bonito. Chaco Canyon, New Mexico. Bill Davis.

p. 11. Sun Breaking through Clouds. New Mexico. Bill Davis.

p. 14. Picurius Moon. New Mexico. Bill Davis.

p. 16. Hot Cloud. New Mexico. Bill Davis.

CHAPTER TWO. Valley of the Gods. Southern Utah. Bill Davis.

p. 23. Galaxy. Stream. Taos, New Mexico. Bill Davis.

p. 26. Tree of Light. Valley of the Gods. Utah. Bill Davis.

p. 31. Herkimer Crystal. Michael Tirvana.

p. 34. High Country Aspens. New Mexico. Bill Davis.

p. 37. Wall Panel from the Palace of Ashur-Nasir-Apal II at Nimrud. Winged being kneeling beside the sacred tree. Metropolitan Museum of Art (885-860 B.C.).

p. 42. Valley Moon. Taos, New Mexico. Bill Davis.

p. 45. Autumn Shows. Rio Grande Gorge. Taos, New Mexico. Bill Davis.

CHAPTER THREE. Cathedral Rock. Yosemite. C.E. Watkins. Metropolitan Museum of Art.

p. 52. Mountain Shadows. Taos, New Mexico. Bill Davis.

p. 59. Sipapu Bridge. Natural Bridges Park, Utah. Bill Davis.

p. 62. Pueblo Bonito Doorways. Chaco Canyon, New Mexico. Bill Davis.

p. 64. Mt. Kailasa. Rommel and Sadhana Varma. The Himalaya Kailasa-Manasarova. Lotus Books. Switzerland.

p. 66. Canyon Light. Rio Pueblo Gorge. Taos, New Mexico. Bill Davis.
p. 69. Chichen Itza. Hunbatz Men.
p. 72-73. Cathedral Rock. Yosemite. C.E. Watkins. Metropolitan Museum of Art.

BIBLIOGRAPHY

William Bloom and Marko Pagacnik. *Leylines and Ecology.* Gothic Image Publications. 1986.

Moyra Caldecott. "Guardians of the Tall Stones." *The Sacred Stones Trilogy.* Celestial Arts, Berkeley, CA. 1986.

Joseph Campbell. *The Mythic Image.* Bollingen Series C. Princeton University Press. 1974.

Christine Downing. *Goddess: Mythological Images of the Feminine.* Crossroad Publishing Co. 1981.

Tom Graves. *Needles of the Stone Revisited.* Gothic Image Publications. 1986. Galstonbury, Somerset, England.

Hesiod. Translated by Richard Latimore. University of Michigan Press. Ann Arbor, MI.

Richard Leviton. "Anti Gravity and the World Grid." *Ley Lines and the Meaning of Adam.* Adventures Unlimited Press. Stelle, IL.

Sig Lonegren. *Spiritual Dowsing.* Gothic Image Publications. Glastonbury. 1986.

Norman Meyers. *The Primary Source.* W.W. Norton & Co., 1984.

John Michell. *The New View over Atlantis.* Harper & Row. 1983.

G. Reichel-Dolmatoff. "Cosmology as Ecological Analysis: A View from the Rain Forest." Huxley Memorial Lecture. Royal Anthropological Institute. 1975.

Michael Roads. *Talking with Nature.* H.J. Kramer, Inc. Tiburon, CA. 1987.

Monica Sjoo and Barbara Mor. *The Great Cosmic Mother.* Harper & Row. 1987.

Rommel and Sadhana Varma. *The Himalaya Kailasa — Manasarova.* Lotus Books. Switzerland.

INDEX

Air p. 44
Alignment p. 5, 7, 8, 17, 18
Amazon River p. 35
Amazon Rain Forest p. 35, 53
Angels p. 27
Ashville, NC p. 60
Australian Aboriginals p. 57
Avebury p. 60
Avicenna p. 39
Ayers Rock p. 58

Bisbee, Arizona p. 60
Body p. 39
Bodhi Tree p. 33

Celebration p. 12
Celts p. 52
Ceremony p. 12
Chichen Itza p. 68, 69
Chimayo, New Mexico p. 40
Consecration p. 9, 10
Cosmic p. 9
Crystal(s) p. 5, 29, 30, 41, 60
Crystalline Grid p. 53, 55, 57, 58

Delphi p. 21
Deva, Devic Kingdom p. 25
Desana Indians p. 36, 38

Earth p. 1, 8, 10, 18, 24, 30, 39, 40, 41, 43, 46, 49, 50, 52, 53, 54, 55, 56, 58, 65, 67
Ecology p. 36, 38
Elements p. 43
Elen Helena p. 52
Endocrine p. 41
Energy, p. 50, 53, 55
   Energy currents p. 49, 50, 52
Ether p. 44
Etheric p. 50, 53

Fire p. 44
Forest p. 33, 34

Gaea p. 21, 52
Galaxy(ies) p. 7, 8, 15
Galen p. 39
Glastonbury p. 60

Grid p. 52, 53, 57, 58, 60
Grid Pattern p. 15, 52, 55, 56, 60

Haleakala p. 60
Healing p. 39, 49
Heavens p. 9, 12, 18
Hesiod p. 21
Holy Mountains p. 61
Horticulture p. 28, 29
Hot Springs, Arkansas p. 21

Kabbalah p. 33
Kneipp p. 39

Land p. 30
Law,
   Law of Reciprocity p. 2, 4
Ley Lines p. 29, 52
Light p. 1, 5, 10, 24, 46, 53, 55

Machu Picchu p. 5, 41
Mayan p. 35, 65, 68
Meditation p. 18, 46, 56, 65
Mining p. 29, 32, 58
Moon p. 13, 15, 18, 68
Mountains (Holy),
   Mt. Athos p. 63
   Mt. Kailasa p. 63, 65
   Mt. Sinai p. 63
   Mt. Tabor p. 63
   Taos Mt. p. 23, 42, 45, 52, 63, 66

Nature p. 22, 24, 25, 27, 29
Nature Spirits p. 25
Newgrange p. 5, 60
Nourishment p. 40, 42

Physical p. 39
Pliny the Elder p. 39
Power Centers p. 28, 55
Purification p. 2

Rainbow Serpent p. 57
Reciprocity p. 2, 4, 49
Revelations p. 27

Sacred p. 32, 56
Sacred Sites p. 58, 60, 61, 65, 68
Shaman p. 36, 38

Solar System p. 7, 68, 70
Stars p. 13, 15
Stonehenge p. 5, 52, 60
Structures p. 8, 55
Subtle p. 7
Sun p. 5, 7, 13, 18

Tambopata Reserve p. 35
Theogony p. 21
Tree of Life p. 33
Tree of Jesse p. 33
Tribal Ecology p. 36, 38
Tukano Indians p. 36, 38

Vibration p. 4, 65
Vision p. 12, 28

Water p. 39, 43
Wullumbum p. 58

Uxmal p. 68

Yucatan p. 60, 67

Zaggurats p. 61
Zohar p. 33